Cambridge Elements ≡

Elements of Paleontology

UTILIZING THE PALEOBIOLOGY DATABASE TO PROVIDE EDUCATIONAL OPPORTUNITIES FOR UNDERGRADUATES

Rowan Lockwood
William and Mary

Phoebe A. Cohen
Williams College

Mark D. Uhen
George Mason University

Katherine Ryker
University of South Carolina

Paleontological
S O C I E T Y

CAMBRIDGE
UNIVERSITY PRESS

CAMBRIDGE
UNIVERSITY PRESS

University Printing House, Cambridge CB2 8BS, United Kingdom

One Liberty Plaza, 20th Floor, New York, NY 10006, USA

477 Williamstown Road, Port Melbourne, VIC 3207, Australia

314–321, 3rd Floor, Plot 3, Splendor Forum, Jasola District Centre,
New Delhi – 110025, India

79 Anson Road, #06–04/06, Singapore 079906

Cambridge University Press is part of the University of Cambridge.

It furthers the University's mission by disseminating knowledge in the pursuit of
education, learning, and research at the highest international levels of excellence.

www.cambridge.org
Information on this title: www.cambridge.org/9781108717908
DOI: 10.1017/9781108681667

First published 2018

A catalogue record for this publication is available from the British Library.

ISBN 978-1-108-71790-8 Paperback
ISSN 2517-780X (online)
ISSN 2517-7796 (print)

Utilizing the Paleobiology Database to Provide Educational Opportunities for Undergraduates

Elements of Paleontology

DOI: 10.1017/9781108681667
First published online: October 2018

Rowan Lockwood
William and Mary

Phoebe A. Cohen
Williams College

Mark D. Uhen
George Mason University

Katherine Ryker
University of South Carolina

Abstract: Integration of research experiences into the undergraduate classroom can result in increased recruitment, retention, and motivation of science students. "Big data" science initiatives, such as the Paleobiology Database (PBDB), can provide inexpensive and accessible research opportunities. Here, we provide an introduction to what the PBDB is, how to use it, how it can be deployed in introductory and advanced courses, and examples of how it has been used in undergraduate research. The PBDB aims to provide information on all fossil organisms, across the tree of life, around the world, and through all of geologic time. The PBDB Resource Page (paleobiodb.org/#/resources) contains a range of PBDB tutorials and activities for use in physical geology, historical geology, paleontology, sedimentology, and stratigraphy courses. As two-year colleges, universities, and distance-based learning initiatives seek research-based alternatives to traditional lab exercises, the PBDB can provide opportunities for hands-on science activities.

Keywords: big data, database, paleobiodiversity, paleobiogeography, authentic research

ISBNs: 9781108717908 (PB), 9781108681667 (OC)
ISSNs: 2517-780X (online), 2517-7796 (print)

Contents

1 Introduction

A number of studies have emphasized potential links between research experience and recruitment and retention of undergraduate science majors (e.g., Nagda et al., 1998; Lopatto, 2009; Singer et al., 2012). However, implementation of such approaches is often hampered by lack of lab space and curricular materials. While many two-year and four-year colleges lack extensive fossil, mineral, and rock collections, all have access to big data science initiatives, such as the Paleobiology Database (PBDB, paleobiodb.org). Although this database is used extensively by researchers (over 1000 scientists from 31 countries), its incredible potential as a learning tool – for undergraduates in particular – has yet to be fully realized.

Early engagement in authentic research opportunities in the classroom has been linked to increased recruitment, more positive scientific attitudes and values, greater retention and persistence, and increased motivation of undergraduate students in Science, Technology, Engineering, and Mathematics (STEM) disciplines (e.g., Nagda et al., 1998; Lopatto, 2009; Singer et al., 2012). The evidence in support of research-based approaches has compelled the President's Council of Advisors on Science, Technology, Engineering, and Mathematics (PCAST) to advocate for the elimination of standard laboratory courses in favor of discovery-based courses (Olson and Riordan, 2012). As the cyberinfrastructural framework of the geosciences evolves (for example EarthChem, GEOROC, PETDB, NAVDAT; see also EarthCube for an overview of NSF [National Science Foundation]-funded cyberinfrastructure projects in geosciences; earthcube.org), the extent to which these resources are used in the undergraduate classroom lags far behind (Lehnert et al., 2006; Ratajeski, 2006). Research databases may provide an effective research tool that is inexpensive, accessible, and easily disseminated to two- (2YC) and four-year college (4YC) introductory geology courses, including distance learning initiatives (Teaching with Data, Simulations, and Models, 2014).

1.1 What Is the Paleobiology Database?

The PBDB is a large open access database that seeks to catalogue all fossil collections and occurrences, through geologic time, and across the tree of life to enable data-driven projects that ask the big questions in paleontology. PBDB data types consist of taxonomic names (including classifications, synonymies, and re-identifications); geographic information (including county, state, province, country, latitude, longitude, and paleo-coordinates calculated to take into account plate tectonic movement); stratigraphic information (including stratigraphic group, formation, member); bibliographic information; and geologic time scale (Uhen et al., 2013). These data are linked together as occurrences—in which a taxon (e.g., species or genus) is recorded from a particular place (geographic information) and a particular time (stratigraphic information, geologic time scale) by a particular author (bibliographic information). These basic pieces of information are sometimes supplemented by paleoecological (e.g., abundance, life habit, feeding mode), geologic (e.g., lithology), paleoenvironmental, sedimentological, morphological (e.g., measurements of size and shape), sampling (e.g., mode of collection), and preservational (e.g., mode of preservation, taphonomic damage) information. Primary literature sources represent the majority of the data, but legacy databases and unpublished data can be contributed for active research projects. As of May 2018, the PBDB contains data for over 65,000 references, 370,000 taxa, and 1,369,000 occurrences in 193,000 collections. These data were entered by 410 users from 130 institutions in 33 countries.

Scientific output from the PBDB has been truly remarkable. As of May 2018, the list of official PBDB publications numbered 311 (PBDB Publications, 2018) with over 872 publications total using PBDB data in some fashion (Google Scholar, 2018). Citations to these papers exceed 27,000, giving the PDBD a Google Scholar h-index of 74. These data have been used for the original purpose of the database, which was to quantify patterns of paleobiodiversity on Earth (Alroy et al., 2001, 2008; Tennant et al., 2018). Applications have extended well beyond this to include conservation paleobiology (Barnosky et al., 2011; Harnik et al., 2012; Blois et al., 2013; Finnegan et al., 2015), biases in the fossil record or lack thereof (Foote, 2001; Uhen and Pyenson, 2007), extinction (Payne and Finnegan, 2007; Button et al., 2017), taphonomy (Behrensmeyer et al., 2005; Kosnik et al., 2011), taxonomy (Soul and Friedman, 2015), the history of paleontological discovery (Uhen, 2010; Uhen et al., 2013), and more recently, the use of the PBDB as an educational tool (George et al., 2016; Lockwood et al., 2016; Bentley et al., 2017; Lukes et al., 2017; Ryker et al., 2017).

1.2 Big Data's Role in Undergraduate Education

The recognition that teaching the practice of science is more important than memorizing the product of science has been reinforced by policy statements across STEM disciplines. The Discipline-Based Education Report (DBER) report from the National Research Council (Singer et al., 2012) clearly states that research-based instructional strategies are more effective in developing conceptual knowledge and improving attitudes about learning than traditional lecture formats. In the geosciences, the critical importance of field and research-based experiences to geoscience major recruitment and retention has long been recognized (Houlton, 2010), but has proven difficult to implement, due to the expense and inaccessibility of research tools. These problems are particularly acute at two-year institutions, which may lack the research infrastructure and facilities to promote "doing" in entry-level or distance-learning courses.

The PBDB, like other research databases, can provide an accessible and inexpensive research tool applicable to entry-level courses at both two- and four-year colleges. Countless STEM databases exist, and visualizations from a limited number of these databases have been leveraged as conceptual demonstrations or parts of inquiry labs (see for example Science on a Sphere by National Oceanic and Atmospheric Administration [NOAA]). How the use of these databases has increased student access to research experiences has received remarkably little attention in geoscience or other STEM disciplines (Singer et al., 2012). This is despite the fact that, during the 2014 Future of Undergraduate Geoscience Education Summit, the geoscience community identified student skills working with big data and datasets as a critical, often missing, component of an undergraduate geoscience education (Mosher et al., 2014). Similarly, the DBER report from National Research Council (Singer et al., 2012) emphasized that inquiry-based student activities using online datasets have been found to improve student learning (e.g., Rissing and Cogan, 2009), increase student competency with science practices (e.g., Brickman et al., 2009), and increase student confidence in their ability to do science (e.g., Seymour et al., 2004). As far back as 2003, geoscience educators explicitly called for the development of database-specific research skills (Manduca and Mogk, 2003), but progress has been limited. As geoscientists transition from traditional lab to authentic research courses, the need for easily accessible sources of big data will only increase.

There are numerous benefits to teaching earth science in the context of cyberinfrastructure, including:

1. Facilitation of effective teaching pedagogies and student skills such as inquiry and evidence-based decision making and analysis (Teaching with Cyberinfrastructure, 2014).
2. Production of customized datasets relating to geologic time, plate tectonics, climate change, mass extinction, and geospatial data.
3. Highlighting the scale and diversity of materials and processes in the Earth system (Teaching with Cyberinfrastructure, 2014).

4. Development of quantitative skills including data transformation, univariate and multivariate statistics, calculation of rates, graph building and interpretation.

5. Introduction to research skills including finding and compiling primary literature, critical thinking, developing hypotheses, working independently, problem solving, data manipulation, etc. (Kardash, 2000; Ishiyama, 2002).

6. Building of technological familiarity literacy for students and instructors.

7. Socialization of students into scientific culture (Hunter et al., 2006).

8. Appreciation of the community and team-based nature of modern science (Understanding Science, 2014). Pedagogical research emphasizes the utility of group learning, but many undergraduates view this approach with skepticism, having few examples of collaborative science to build on.

9. Facilitation of communication and collaboration between two- and four-year college geology departments without significant travel between them.

2 How to Use the Paleobiology Database

2.1 How to Access the PBDB

The PBDB is organized as a relational database with multiple web forms for data entry, visualization, and retrieval. In addition, an Application Programming Interface (API) has been developed, which simplifies data retrieval tasks and makes it much easier for scientific and/or educational users to develop their own web or mobile-based applications (apps) to search, download, visualize, and analyze data. A data input API is also under development that will allow the creation of customized data input applications as well. A number of PBDB apps currently exist and are accessible from the PBDB Resources webpage (paleobiodb.org/#/resources). The two most commonly used applications for searching, visualizing, and analyzing PBDB data are PBDB Navigator and Fossilworks. PBDB Navigator (paleobiodb.org/naviga tor/) was developed by Shanan Peters, Michael McClennen, and John Czaplewski and launched in 2013. Navigator allows users to explore when and where fossils occur, in addition to mapping their geographic distribution, graphing their diversity through time, and examining their taxonomy. Information on how to use PBDB Navigator is provided below. Fossilworks (fossilworks.org/), which was launched by John Alroy in 2013 based on the original Paleobiology Database platform, allows users to search all aspects of the database and map occurrences. Downloading PBDB data, for users to analyze on their own, can be accomplished using the PBDB Application

Programming interface (API; paleobiodb.org/data1.2/), PBDB website (paleo biodb.org/classic/displayDownloadGenerator), or Fossilworks.

The PBDB seeks to catalogue all fossil collections and occurrences – through geologic time, around the globe, and across the tree of life. The primary record for most fossils in the PBDB is the occurrence, which is defined as where an individual sample or set of samples of a fossil species occurs in space (geographically) and time (stratigraphically). For example, 10 specimens of a species of trilobite found in a rock unit dated to the late Cambrian would be an occurrence of that taxon. A PBDB collection is a set of occurrences from the same site (i.e., locality). For example, a collection might be a group of eight different fossil taxa occurrences (which themselves could constitute one or more actual specimens of the same taxon) found in one outcrop. One of the easiest ways to access the PBDB is using a web application called PBDB Navigator (paleobiodb.org/navigator/).

2.2 Delving into PBDB Navigator

PBDB Navigator consists of five parts (Fig. 1):

1. Map (center of screen, Fig. 1) showing continents with dots representing fossil collections. The color of these dots represents their geologic age. If you zoom in and click on the dots, you can see the information on each site and the taxa that occur there. The map changes as you zoom in or out, search for specific taxonomic names, and/or select particular time intervals (see below).
2. Geologic Time Scale (bottom of screen, Fig. 1) showing the major eras, periods, and stages. If you click on the geologic time scale, the map will

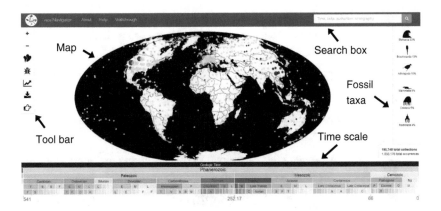

Figure 1. PBDB Navigator "landing page" labeling the map, toolbar, geologic time scale, search box, and fossil taxa bar

show you the location of all fossil collections from that time interval. For broader time resolution, you can click on the upper rows of the geologic time scale, including eon (Phanerozoic) or era (e.g., Paleozoic, Mesozoic, Cenozoic). For more specific time resolution, you can select the lower rows of the geologic time scale, including periods (e.g., Jurassic, Cretaceous) and stages (e.g., Aptian, Albian). Double clicking on a time period will zoom the time scale to that particular time period.

3. Tool Bar (left side of screen, Fig. 1) showing the tools you can use to explore the database (Table 1). If you position the mouse cursor over each tool, the tool name will pop up. The *Zoom in* and *Zoom out* tools allow you to zoom in and zoom out of the map, so that you can focus on a particular geographic area. The *Toggle Paleogeography* button allows you to switch between modern and paleo tectonic plate configurations, but note that you must select a period or lower time interval first. The *Toggle Taxa* button allows you to open a taxonomic browser, which you can use to search for particular taxonomic names at any taxonomic level. As you type in this browser, the taxonomic name will be auto-filled with the name of organisms contained in the PBDB. The *Toggle Stats* button makes it possible to quickly and easily create diversity, extinction, and origination curves. It is important to note that the data for calculating diversity, extinction, and origination

Table 1 Toolbar for PBDB Navigator

Symbol	Description
✚	*Zoom in/out* on the map
🝖	*Toggle Paleogeography*: Reconstructs plate tectonic configurations for time interval (era or smaller) you are exploring
🐜	*Toggle Taxa* Browser: Narrow down which taxonomic group is plotted on map
📈	*Toggle Stats*: Create a diversity curve for the collections currently plotted on map
⬇	*Save Map Data*: Download the data (lat/long, geologic age, etc.) for the occurrences plotted on map
👉	*See Examples*: Start using PBDB Navigator with three examples

curves is determined by the geographic region that is currently displayed on the map. Finally, the *Save Map Data* button allows you to download the data for the occurrences plotted on the map.

4. Search Box (upper right corner of screen, Fig. 1) allowing you to search for particular time intervals, taxa, or stratigraphic unit. As you type in the search box, the words will be auto-filled with the time intervals, taxonomic names, and stratigraphic units contained in the PBDB.

5. Fossil Taxa Bar (right side of screen, Fig. 1) showing the faunal composition (in percent) of the collections currently represented on the map, plus the total number of collections and occurrences. This information changes as you zoom in or out, search for specific taxonomic names, and/or select particular time intervals.

A video tutorial (five minutes long) on how to use PBDB Navigator is available on the PBDB Resources page (paleobiodb.org/#/resources), along with other useful tutorials.

3 Educational Integration: Examples and Ideas

Here we present a handful of data-focused lessons for undergraduate geoscience courses using the PBDB. These lessons are a small subset of the lessons and tutorials developed as part of an NSF-funded study on the use of research databases in undergraduate education (NSF IUSE grant DUE-1504588 and 1504718). These resources are available on both the PBDB Resources webpage (paleobiodb.org/#/resources) and the SERC (Science Education Resource Center, serc.carleton.edu) educational resource platform (search term PBDB). Each lesson has undergone peer-review and been piloted at a variety of two- and four-year institutions, including George Mason University (Virginia), William and Mary (Virginia), Eastern Michigan University (Michigan), High Point University (North Carolina), Northern Virginia Community College (Virginia), and Thomas Nelson Community College (Virginia).

These lessons have been designed to: 1) be modular so that they range in duration from five minutes to three hours and 2) be flexible for use in lecture, laboratory, and field settings, or as homework. The lessons feature essential skills for scientifically literate citizens (including critical thinking, hypothesis generation, and data analysis; see Next Generation Science Standards) applied to a range of societally important topics, including extinction, evolution, and climate change.

A note on access and accessibility: PBDB Navigator requires the use of a laptop or desktop computer and Internet access. While many institutions are able to provide access to computing facilities, some are not, and emphasizing economic disparities among students and/or institutions should be avoided.

3.1 Integration into Introductory-Level Classes

The ease with which undergraduate students can explore PBDB Navigator makes it ideal for use in introductory courses, including physical geology, physical geography, and historical geology. It is a quick, inexpensive, and easy way to introduce scientific data into undergraduate teaching, which does not require substantial lab or research facilities. This makes it particularly effective for use by two-year colleges and distance learning initiatives.

As a quick five-minute activity during lecture or lab, instructors could ask students to look up which fossils or fossil sites occur in their local area (e.g., university location, hometown, etc.). As a longer-term lab or homework activity (20–40 minutes), students could determine which fossils occur through time in a particular region (state, province), then hypothesize how paleoenvironment has changed through time based on these fossils. This is a particularly effective exercise if the instructor chooses a region that has experienced substantial changes in environment (e.g., from marine to terrestrial) through time.

3.1.1 Pangea Puzzle

One activity, which allows instructors to emphasize the general importance of paleontology at the introductory level, focuses on the distribution of fossils across the supercontinent of Pangea. The Pangea Puzzle activity is accessible via the PBDB Resource Page or the SERC Resource Platform (serc.carleton .edu/NAGTWorkshops/intro/activities/177872.html), and includes learning objectives, classroom handouts, teaching tips, and suggestions for assessment. This activity can be used in an introductory or intermediate undergraduate course, including (but not limited to) physical geology, historical geology, and paleontology. Students can work as individuals or in pairs, and class size can range from a small seminar (< 10 students) to a large lecture (> 100), as long as sufficient computer facilities are available. It has been successfully piloted with non-majors and majors from both two- and four-year institutions. Each student or student pair will need access to a laptop or desktop computer connected to the Internet, running an Internet browser. The activity can be used in its entirety as a lab or homework activity (1–1.5 hours) or pared down to be used as a lecture activity (20–30 minutes). It is most effectively presented as a supplement to coursework on Pangea, continental drift, and plate tectonics.

In this activity, students learn how to use the PBDB to map the geographic distribution of several fossil taxa, using both modern and past plate configurations. They then use these maps to explore how the geographic distribution of fossils provides evidence for the supercontinent Pangea and plate tectonic movement. Concept goals for this activity include: 1) constructing a map of fossil occurrences on the present-day Earth's surface, 2) constructing a map of fossil occurrences on the Earth's surface at various times in Earth's past, 3) identifying the past distributions of fossils on ancient continents and supercontinents, and 4) explaining how the present-day distributions of fossil organisms are different from their distribution during the time of their deposition as fossils. A higher-order learning objective involves developing hypotheses regarding why the present-day distribution of fossil occurrences is dramatically different from their distribution in the past.

To begin the activity, students are asked to map the geographic distribution of a Permian fossil taxon, such as *Glossopteris* (seed fern), on a map with modern tectonic plate configurations. *Glossopteris* works particularly well for this activity, but students can use several taxa, including the original examples used by Alfred Wegener to support his ideas on continental drift, including *Lystrosaurus* (dicynodont therapsid, transitional fossil between reptiles and mammals) and *Mesosaurus* (aquatic reptile). Students can then be asked to figure out what each organism is (using the *Toggle Taxa* button on the left tool bar), when their fossils occur (using the geologic time scale at the bottom of the screen), and where they occur (using the map in the center). If students limit the distribution of mapped occurrences to the Permian (by clicking on the Permian interval on the geologic time scale at the bottom of the screen), they'll see that, for example, the distribution of *Glossopteris* is limited to the southern continents—South America, Africa, Australia, and southern Asia (Fig. 2A). The instructor can then reveal that these taxa cannot travel long distances over water, and ask students to explain why this is a problem, given the modern distribution of these taxa.

Next, the students can click on the *Toggle Paleogeography* button (on the left tool bar) to map the distribution of Permian *Glossopteris* on a map with Permian plate configurations (Fig. 2B). Students can then be asked to explain how this distribution provides evidence of the supercontinent Pangea and plate tectonics. Follow-up questions can focus on: 1) what other types of taxa could be used to reconstruct plate positions through time, 2) what direction each continent is moving through time, and 3) what the estimated rate of spreading is for the Atlantic Ocean (if instructor provides additional information regarding the current size of the Atlantic).

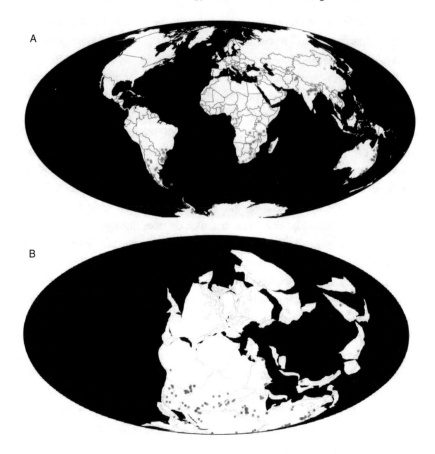

Figure 2. Example of output from Puzzling Pangaea activity using PBDB Navigator. (A) Students map the distribution of fossil occurrences of *Glossopteris* ferns during the Permian, superimposed on modern continental configuration. (B) After attempting to explain this distribution, students then map the same data, overlain on continental configuration for the continent of Pangaea during the Permian.

The web link for this activity (serc.carleton.edu/NAGTWorkshops/intro/activities/177872.html) provides a wealth of information including approaches for student assessment, background reading, and pre-requisite skills.

3.2 Integration into Upper-Level Classes

The wealth of fossil data contained in the PBDB makes it very useful for upper-level courses, including paleontology, sedimentology, and stratigraphy. Quick activities (10 minutes or less) could include mapping the distribution of a particular fossil group across both space and/or geologic time; looking up the taxonomic classification of a fossil organism; and exploring the local fossil record to identify nearby localities and obtain faunal lists. Longer activities (30–60 minutes) with a paleontological focus could range from plotting diversity, extinction, and origination curves through time; to exploring the effects of events such as the end-Permian mass extinction and the Great American Biotic Interchange on biodiversity; to quantifying the magnitude of major mass extinctions; to tracking climate change through time and space using fossil proxies of temperature or rainfall. PBDB data also lend themselves to use in more sedimentary or stratigraphy focused courses, for example, through tracking sea level or depositional environment through time using fossil proxies for paleoenvironment; to downloading data for biostratigraphic and biogeographic analyses. The PBDB can also facilitate longer-term lab activity and class projects by providing instructors with large amounts of data that they can use for teaching students how to apply a variety of quantitative techniques (e.g., extinction and origination rates, sample standardized metrics of diversity, rarefaction, ordination techniques, survivorship curves, etc.).

3.2.1 Counting Critters

Recent enhancements to the PBDB toolbar have made it possible to quickly and easily calculate diversity, extinction, and origination rates. The Counting Critters activity takes advantage of these new tools and can be accessed via the PBDB Resource Page and the SERC Resource Platform (serc.carleton.edu /NAGTWorkshops/intro/activities/185138.html). On this website, you will find learning objectives, background information, classroom handouts, teaching tips, and suggestions for assessment. This activity is designed for use in an intermediate or advanced course, such as paleontology and historical geology, as a lecture, lab, or homework activity. Class size can range from small (< 10) to medium (< 50), as long as sufficient computer facilities (one computer per 1–3 students) and teaching assistants are available. This activity has been piloted with majors from a four-year institution. It can be used in its entirety as a lab or homework activity (1.5 hours) or divided into modules for use in lecture (20–30 minutes). The activity is most effective when presented as part of a unit on diversity through time, extinction, and/or origination.

For this activity, students learn how to use PBDB Navigator to develop a diversity curve showing changes in global biodiversity through time. They then use this curve to explore major events in the history of life, including the Cambrian Explosion and the end-Permian extinction. As an optional supplement, students can explore the effects of sampling and preservation on diversity curves. After completing this activity, students will be able to: 1) construct a diversity curve using data and tools from PBDB Navigator, 2) interpret graphical representations of diversity curves to identify possible increases and decreases in diversity, 3) identify a major origination (Cambrian Explosion) graphically and use Internet sources to research its possible causes, 4) identify a major extinction (end-Permian extinction) graphically and use Internet sources to research a possible cause, 5) assess the effects of sampling and preservation on quantifying diversity, 6) assess the extent to which diversity patterns are affected by the inclusion of singleton taxa, and 7) determine the extent to which Pull of the Recent is influencing diversity patterns.

Before students delve into PBDB Navigator, instructors can build their higher-order thinking skills by asking them to brainstorm whether they expect global diversity to increase, decrease, or stay the same through the past 500 million years. Next, students can use the PBDB to calculate a diversity curve – for all organisms, globally, throughout the past 500 million years (Fig. 3). To do this, they zoom the PBDB map out as far as possible. Then, they click on the *Toggle Stats* button (on the left toolbar) and

- For taxonomic level, select Genus
- For temporal resolution, select Age
- Then click on ***Use Advanced Diversity Curve Generator*** button
- WAIT. Students need to be patient; it can take 5–10 minutes to calculate all of life's diversity through time!
- When a graph appears on the screen, check the box *Range through Diversity*
- Uncheck the boxes ***Sampled-in-bin Diversity*** and ***Include Singletons***

These particular instructions will produce a range-through diversity curve, which assumes that each taxon is present from its first occurrence to its last occurrence, whether or not it was sampled in the fossil record in the intervening intervals (Fig. 3A). Students can also choose to plot sampled-in-bin diversity, which is the number of taxa actually found in each time interval. Singletons, or taxa that occur in only one time interval, can be included or excluded. Once students have produced a curve, instructors can ask their students to describe the curve, including the general trend through time. If different students are obtaining different graphs, it is most likely because they: 1) did not zoom out to obtain a global map and/or 2) they have selected a particular time interval, taxonomic name, or stratigraphic unit in the search box. Students can identify rapid increases in diversity, including the Cambrian Explosion, Ordovician Radiation, and the recovery after the end-Permian extinction. Several rapid decreases in diversity can be recognized, such as the end-Permian, end-Triassic, end-Cretaceous, and end-Eocene extinction events. Once students have

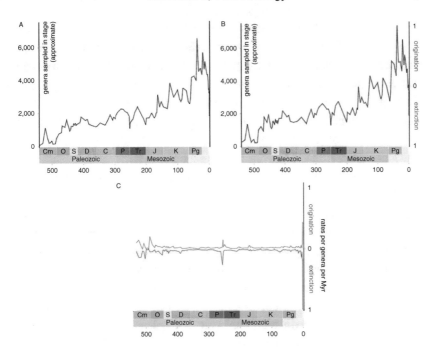

Figure 3. Samples of output from Counting Critters activity using PBDB Navigator. (A) Students reconstruct a diversity curve through time, for all fossils, around the world during the past 500 million years. They can then use this diversity curve to explore increases (e.g., Cambrian Explosion, Ordovician Radiation) and decreases (e.g., end-Permian, end-Cretaceous extinctions) in diversity. Students can expand on this and explore the effects of excluding (A) versus including (B) singletons, which are organisms that occur in only one time interval. (C) PBDB Navigator also makes it possible to calculate per capita origination and extinction rates (Foote 2000).

produced a diversity curve, they can save the image (Fig. 3A) and download the diversity data for subsequent analyses. The downloaded data can be easily opened in Excel, R, and other programs, and include geological stage names, minimum and maximum age for each stage (millions of years ago), and a diversity estimate. Students can use these data to estimate the magnitude of some of the events they identify on their curves.

Instructors can explain the concepts of sampling and preservation to students and ask them a series of questions to brainstorm how these factors might influence estimates of diversity through time. If the activity is presented towards the end of a paleontology course, students could predict which time intervals are likely to be over- or under-sampled and which taxa are likely to be well or poorly preserved. Several studies have suggested that the number of singletons in a time interval should increase when sampling and preservational quality decrease (Pease, 1985; Alroy, 1998; Foote, 2000). Students can explore this concept by comparing the diversity curve excluding (Fig. 3A) and including (Fig. 3B) singletons. Instructors can also discuss the concept of Pull of the Recent, the extent to which it affects the diversity curve, and techniques for quantitatively tackling the issue (Raup 1972, 1976, 1979; Sepkoski, 1997; Peters and Foote, 2001; Jablonski et al., 2003; Alroy et al., 2008).

Moving beyond the *Counting Critters* activity, instructors can also use PBDB Navigator to explore extinction and origination rates. Origination and extinction rates are estimated in PBDB Navigator using the per capita rate method of Foote (2000). This method compares the number of taxa that exist both before and after each interval to those that either originate or go extinct during it. Students can be asked to plot, save, and interpret both origination and extinction rates (Fig. 3 C). The origination rate curve makes it easy to spot several events – including the Cambrian Explosion, Ordovician Radiation, and recovery after the end-Permian. The curve for extinction rates emphasizes the end-Permian extinction above any other event, but students can also document the end-Ordovician, late Devonian, and end-Cretaceous events (Fig. 3 C). Note that many methods exist for calculating diversity and origination/extinction rates. Those provided by PBDB Navigator are among the simplest for undergraduates to explore and understand, but instructors can also apply a series of more complex approaches (e.g., three-timer, boundary-crosser, evenness, dominance, sampling quorum) using Fossilworks. Step-by-step instructions on how to calculate diversity, extinction, and origination curves using Fossilworks are provided on the PBDB Resource Page (under Educational Tutorials, paleobiodb.org/#/resources). Instructors can also take advantage of the download tools in Fossilworks and the PBDB Downloader (paleobiodb.org/classic/displayDownloadGenerator) to download the data for further analysis using R or other software options.

3.3 Integration into Independent Student Research

The incredible diversity of data available in the PBDB lends itself to facilitating independent undergraduate research projects, across a variety of institution types. The PBDB, and other scientific databases, can bring research opportunities to underserved populations, particularly those that participate in distance learning or do not have ready access to lab facilities. Project topics can focus on diversity, extinction/origination rates, biostratigraphy, paleobiogeography, paleoecology, paleoenvironmental reconstruction, and many others. Projects could be developed as short (two-week) initiatives or longer-term (multi-year) approaches, for introductory through advanced undergraduates.

To illustrate how the PBDB can be used in undergraduate research, two case studies are described in detail. The first is a project led by Kristina Okamoto (senior undergraduate at University of California, Santa Cruz), in collaboration with her faculty advisor, Dr. Matthew Clapham. They used the PBDB to test the hypothesis that extinction of ammonoids across the Cretaceous/Paleogene (K/Pg) boundary released competitive pressure in nautilids and drove shifts in morphology and/or increases in abundance. Relying on measurement and occurrence data from the PBDB, they tested for changes in nautilid size and shape across the K/Pg extinction. They documented stasis in body size, at the same time as punctuated change in sutural complexity, suggesting mixed evidence for competitive release. Their work, which Okamoto presented as a poster at the national Geological Society of America conference in 2015 (Fig. 4A) and as a talk in 2017, is currently being prepared for publication.

Another example of incorporating the PBDB into undergraduate research is a project spearheaded by Professor Richard Butler and Sam Tutin (third-year undergraduate) at the University of Birmingham (UK). The project, which began in September of 2015 and was published in 2017 (Tutin and Butler, 2017), involved a detailed assessment of the quality of the plesiosaur (marine reptile) fossil record. They used the PBDB extensively as a source of data on the stratigraphic occurrence and taxonomy of plesiosaurs, and also as a route into the original literature for each species. They quantified completeness, using two metrics: 1) a character completeness metric (percentage of phylogenetic characters from a cladistic dataset that can be scored for that species) (Fig. 4B) and 2) a skeletal completeness metric (i.e., percentage of the overall skeleton that is preserved for that species). Tutin and Butler (2017) found that the plesiosaur fossil record was statistically significantly more complete than terrestrial records for contemporaneous groups (e.g., sauropods, pterosaurs, birds).

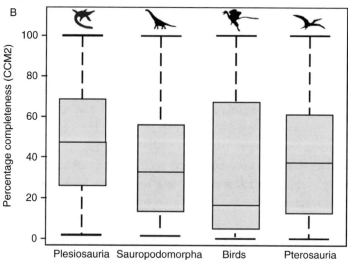

Figure 4. Case studies focusing on undergraduate research using the PBDB. (A) At the University of California, Santa Cruz, Kristina Okamoto and Dr. Matthew Clapham used the PBDB to explore the evolutionary and ecological effects of the K/Pg extinction of ammonoids on nautilid cephalopod mollusks. Okamoto presented their results as a poster at the national Geological Society of America conference in 2015. (B) Sam Tutin and Professor Richard Butler, at the University of Birmingham (UK), used the PBDB to evaluate the

4 Conclusions

Research databases, including the PBDB, can be integrated into undergraduate teaching to provide authentic research experiences in the classroom, lab, and at home. The online accessibility of these databases makes them particularly useful for distance learning and at institutions that have limited laboratory infrastructure or teaching collections. The data contained in the PBDB include which fossil taxa occur when and where – making it useful for covering topics such as local fossils, diversity and extinction, climate change, and sea level change. Activities have been developed for a variety of introductory courses (e.g., physical geology, historical geology) and advanced courses (e.g., paleontology, sedimentology, stratigraphy), ranging in duration from five minutes to three hours. Examples of activities and tutorials are readily available on the PBDB Resource page (paleobiodb.org/#/resources). The PBDB represents an incredible opportunity to engage students in authentic research activities, while connecting them to a broader community of scientists actively exploring the history of life on earth.

Caption for Figure 4. (cont.)

quality of preservation in plesiosaurs, fossil marine reptiles. This figure, which they published in *Acta Palaeontologica Polonica* in 2017, compares the completeness of the plesiosaur fossil record to contemporaneous marine and terrestrial vertebrates.

References

Alroy, J. (1998). Equilibrial diversity dynamics in North American mammals. In M. L. McKinney and J. A. Drake, eds., *Biodiversity Dynamics*. New York: Columbia University Press, pp. 232–287.

Alroy, J., Aberhan, M., Bottjer, D. J., Foote, M., Fürsich, F. T., Harries, P. J., Hendy, A. J., Holland, S. M., Ivany, L. C., Kiessling, W. & Kosnik, M. A. (2008). Phanerozoic trends in the global diversity of marine invertebrates. *Science*, **321**, 97–100.

Alroy, J., Marshall, C. R., Bambach, R. K., Bezusko, K., Foote, M., Fürsich, F. T., Hansen, T. A., Holland, S. M., Ivany, L. C., Jablonski, D. & Jacobs, D. K. (2001). Effects of sampling standardization on estimates of Phanerozoic marine diversification. *Proceedings of the National Academy of Sciences*, **98**, 6261–6266.

Barnosky, A. D., Matzke, N., Tomiya, S., Wogan, G. O., Swartz, B., Quental, T. B., Marshall, C., McGuire, J. L., Lindsey, E. L., Maguire, K. C. & Mersey, B. (2011). Has the Earth's sixth mass extinction already arrived? *Nature*, **471**, 51–57.

Behrensmeyer, A. K., Fürsich, F. T., Gastaldo, R. A., Kidwell, S. M., Kosnik, M. A., Kowalewski, M., Plotnick, R. E., Rogers, R. R. & Alroy, J. (2005). Are the most durable shelly taxa also the most common in the marine fossil record? *Paleobiology*, **31**, 607–623.

Bentley, C., Ryker, K. & Sundell, A. (2017). *Utilizing the Paleobiology Database to Explore Tectonic Events*. Presentation at Earth Educators' Rendezvous. Albuquerque, NM.

Blois, J. L., Zarnetske, P. L., Fitzpatrick, M. C. & Finnegan, S. (2013). Climate change and the past, present, and future of biotic interactions. *Science*, **341**, 499–504.

Brickman, P., Gormally, C., Armstrong, N. & Hallar, B. (2009). Effects of inquiry-based learning on students' science literacy skills and confidence. *International Journal for the Scholarship of Teaching and Learning*, **3**, 1–22.

Button, D. J., Lloyd, G. T., Ezcurra, M. D., & Butler, R. J. (2017). Mass extinctions drove increased global faunal cosmopolitanism on the super-continent Pangaea. *Nature Communications*, **8**(1), 733.

Finnegan, S., Anderson, S. C., Harnik, P. G., Simpson, C. C., Tittensor, D. P., Byrnes, J. E, et al. (2015). Paleontological baselines for evaluating extinction risk in the modern oceans. *Science*, **348**, 567–570.

Foote, M. (2000). Origination and extinction components of taxonomic diversity: general problems. *Paleobiology*, **26**, 74–102.

(2001). Inferring temporal patterns of preservation, origination, and extinction from taxonomic survivorship analysis. *Paleobiology*, **27**, 602–630.

George, C. O., Bentley, C., Berquist, P. J., Lockwood, R., Lukes, L. A., Ryker, K. & Uhen, M. D. (2016). *Utilizing the Paleobiology Database to Provide Hands-On Research Opportunities for Undergraduates*. Presentation at Earth Educators' Rendezvous. Madison, WI.

Google Scholar. (2018). Paleobiology Database Google Scholar page, scholar .google.com/citations?user=PP946k4AAAAJ, accessed 15 May 2018.

Harnik, P. G., Lotze, H. K., Anderson, S. C., Finkel, Z. V., Finnegan, S., Lindberg, D. R., et al. (2012). Extinctions in ancient and modern seas. *Trends in Ecology and Evolution*, **27**, 608–617.

Houlton, H. R. (2010). *Academic provenance: Investigation of pathways that lead students into the geosciences*. Unpublished M.S. dissertation, Purdue University, 166 p.

Hunter, A., Laursen, S. & Seymour, E. (2006). Becoming a scientist: The role of undergraduate research in students' cognitive, personal, and professional development. *Science Education*, **91**, 36–74.

Ishiyama, J. (2002). Does early participation in undergraduate research benefit social science and humanities students? *College Student Journal*, **36** 381–387.

Jablonski, D., Roy, K., Valentine, J. W., Price, R. M. & Anderson, P. S. (2003). The impact of the Pull of the Recent on the history of marine diversity. *Science*, **300**, 1133–1135.

Kardash, C. (2000). Evaluation of an undergraduate research experience: Perceptions of undergraduate interns and their faculty mentors. *Journal of Educational Psychology*, **92**, 191–201.

Kosnik, M. A., Alroy, J., Behrensmeyer, A. K., Fürsich, F. T., Gastaldo, R. A., Kidwell, S. M., Kowalewski, M., et al. (2011). Changes in shell durability of common marine taxa through the Phanerozoic: Evidence for biological rather than taphonomic drivers. *Paleobiology*, **37**, 303–331.

Lehnert, K., Ratajeski, K. & Walker, D. (2006). *Using online igneous geochemical databases for research and teaching: short course handbook*, Geological Society of America Meeting, Philadelphia, October.

Lockwood, R., George, C. O., Bentley, C., Berquist, P. J., Park Boush, L. E., Lukes, L. A., Ryker, K. & Uhen, M. D. (2016). Using geoscience databases to provide authentic research opportunities for undergraduates. *Geological Society of America Abstracts with Programs* **48**, doi: 10.1130/abs/2016AM-286049.

Lopatto, D. (2009). *Science in Solution: The Impact of Undergraduate Research on Student Learning*. Research Corporation for Science Advancement.

Lukes, L. A., Ryker, K., Millsaps, C., Lockwood, R., Uhen, M. D., Bentley, C., Berquist, P. J. & George, C. O. (2017). Student perceptions of using the Paleobiology Database (PBDB) to conduct undergraduate research. *Geological Society of America Abstracts with Programs* **49**, doi: 10.1130/abs/2017AM-301608.

Manduca, C. & Mogk, D. (2003). Using Data in Undergraduate Science Classrooms, d32ogoqmya1dw8.cloudfront.net/files/usingdata/UsingData .pdf, accessed 16 May 2018.

Mosher, S., Bralower, T., Huntoon, J., Lea, P., McConnell, D., Miller, K., Ryan, J., Summa, L., Villalobos, J. & White, J. (2014). *Future of undergraduate geoscience education: Summary report for Summit on Future of Undergraduate Geoscience Education*, www.jsg.utexas.edu /events/files/Future_Undergrad_Geoscience_Summit_report.pdf, accessed 16 May 2018.

Nagda, B. A., Gregerman, S. R., Jonides, J., VonHippel, W. & Lerner, J. S. (1998). Undergraduate student-faculty research partnerships affect student retention. *Review of Higher Education*, **22**, 55–72.

Olson, S. & Riordan, D. G. (2012). *Engage to Excel: Producing One Million Additional College Graduates with Degrees in Science, Technology, Engineering, and Mathematics*. Report to the President. Executive Office of the President, obamawhitehouse.archives.gov/sites/default/files/micro sites/ostp/pcast-engage-to-excel-final_2–25-12.pdf, accessed 16 May 2018.

Payne, J. L. & Finnegan, S. (2007). The effect of geographic range on extinction risk during background and mass extinction. *Proceedings of the National Academy of Sciences*, USA, **104**, 0506–10511.

PBDB Publications. (2018). *Paleobiology Database Publications*, paleobiodb .org/classic/publications?a=publications, accessed 16 May 2018.

Pease, C. M. (1985). Biases in the durations and diversities of fossil taxa. *Paleobiology*, **11**, 272–292.

Peters, S. E. & Foote, M. (2001). Biodiversity in the Phanerozoic: a reinterpretation. *Paleobiology*, **27**, 583–601.

Ratajeski, K. (2006). A digital resource collection supporting the use of Earthchem databases in geoscience courses. *Geological Society of America Abstracts with Programs*, **38**, p. 524. https://gsa.confex.com /gsa/2006AM/webprogram/Paper113697.html

Raup, D. M. (1972). Taxonomic diversity during the Phanerozoic. *Science*, **177**, 1065–1071.

(1976). Species diversity in the Phanerozoic; an interpretation. *Paleobiology*, **2**, 289–297.

(1979). Biases in the fossil record of species and genera. *Bulletin of the Carnegie Museum of Natural History*, **13**, 85–91.

Rissing, S. W. & Cogan, J. G. (2009). Can an inquiry approach improve college student learning in a teaching laboratory? *CBE-Life Sciences Education*, **8**, 55–61.

Ryker, K., Bentley, C. & Uhen, M. D. (2017). *The Pangea Puzzle*. Presentation at Earth Educators' Rendezvous. Albuquerque, NM.

Sepkoski, J. J. (1997). Biodiversity: Past, present, and future. *Journal of Paleontology*, **71**, 533–539.

Seymour, E., Hunter, A. B., Laursen, S. L. & Deantoni, T. (2004). Establishing the benefits of research experiences for undergraduates in the sciences: First findings from a three-year study. *Science Education*, **88**, 493–534.

Singer, S. R., Nielsen, N. R. & Schweingruber, H. A., eds. (2012). *Discipline-based education research: Understanding and improving learning in undergraduate science and engineering*. National Academies Press.

Soul, L. C. & Friedman, M. (2015). Taxonomy and phylogeny can yield comparable results in comparative paleontological analyses. *Systematic Biology*, **64**, 608–620.

Teaching with Cyberinfrastructure: Benefits and Potential Problems, serc .carleton.edu/research_education/cyberinfrastructure/benefits.html, accessed 16 May 2018.

Teaching with Data, Simulations, and Models, serc.carleton.edu /NAGTWorkshops/data_models/index.html, accessed 16 May 2018.

Tennant, J., Chiarenza, A. A., & Baron, M. (2018). How has our knowledge of dinosaur diversity through geologic time changed through research history? *PeerJ* 6:e4417; DOI 10.7717/peerj.4417.

Tutin, S. L. & Butler, R. J. (2017). The completeness of the fossil record of plesiosaurs, marine reptiles from the Mesozoic. *Acta Palaeontologica Polonica*, **62**, 563–573.

Uhen, M. D. (2010). The origin(s) of whales. *Annual Review of Earth and Planetary Sciences*, **38**, 189–219.

Uhen, M. D., Barnosky, A. D., Bills, B., Blois, J., Carrano, M. T., Carrasco, M. A. et al. (2013). From card catalogs to computers: databases in vertebrate paleontology. *Journal of Vertebrate Paleontology*, **33**, 13–28.

Uhen, M. D. & Pyenson, N. D. (2007). Diversity estimates, biases, and historiographic effects: resolving cetacean diversity in the Tertiary. *Palaeontologia Electronica*, **10**, 1–22.

Understanding Science, undsci.berkeley.edu/article/scienceflowchart, accessed 16 May 2018.

Acknowledgments

The authors would like to thank C. Bentley, P. Berquist, and C. George for inquiry activity development, piloting, and data collection support; L. Lukes for inquiry activity design feedback and evaluation; V. Syverson and S. Peters for technical support and programming; and M. Clapham and R. Butler for help developing undergraduate research case studies. Educational databasing was generously provided by SERC. Funding for this project was provided by NSF IUSE grants DUE-1504588 and 1504718.

Cambridge Elements ≡

Elements of Paleontology

Editor-in-Chief

Colin D. Sumrall
University of Tennessee

About the Series

The Elements of Paleontology series is a publishing collaboration between the Paleontological Society and Cambridge University Press. The series covers the full spectrum of topics in paleontology and paleobiology, and related topics in the Earth and life sciences of interest to students and researchers of paleontology.

The Paleontological Society is an international nonprofit organization devoted exclusively to the science of paleontology: invertebrate and vertebrate paleontology, micropaleontology, and paleobotany. The Society's mission is to advance the study of the fossil record through scientific research, education, and advocacy. Its vision is to be a leading global advocate for understanding life's history and evolution. The Society has several membership categories, including regular, amateur/avocational, student, and retired. Members, representing some 40 countries, include professional paleontologists, academicians, science editors, Earth science teachers, museum specialists, undergraduate and graduate students, postdoctoral scholars, and amateur/avocational paleontologists.

Paleontological
S O C I E T Y

Cambridge Elements ≡

Elements of Paleontology

Elements in the Series

These Elements are contributions to the Paleontological Short Course on *Pedagogy and Technology in the Modern Paleontology Classroom* (organized by Phoebe A. Cohen, Rowan Lockwood, and Lisa Boush), convened at the Geological Society of America Annual Meeting in November 2018 (Indianapolis, Indiana USA).

A full series listing is available at: www.cambridge.org/EPLY

Printed in the United States
By Bookmasters